COUNTDOWN TO SPACE

COMETS AND ASTEROIDS—
Ice and Rocks in Space

Michael D. Cole

CUYAHOGA COUNTY
PUBLIC LIBRARY
2111 Snow Road
Parma, Ohio 44134

Enslow Publishers, Inc.
40 Industrial Road   PO Box 38
Box 398   Aldershot
Berkeley Heights, NJ 07922 Hants GU12 6BP
USA   UK
http://www.enslow.com

Copyright © 2003 by Enslow Publishers, Inc.

All rights reserved.

No part of this book may be reproduced by any means without the written permission of the publisher.

Library of Congress Cataloging-in-Publication Data

Cole, Michael D.
  Comets and asteroids : ice and rocks in space / Michael D. Cole.
     p. cm. — (Countdown to space)
  Summary: Explores what comets and asteroids are, how scientists have studied them throughout history, and the effects of space debris on the Earth when it enters our atmosphere.
  Includes bibliographical references and index.
  ISBN 0-7660-1954-3
  1. Comets—Juvenile literature. 2. Asteroids—Juvenile literature. [1. Comets. 2. Asteroids.] I. Title. II. Series.
  QB721 .C63 2002
  523.6—dc21
                                              2002008520

Printed in the United States of America

10 9 8 7 6 5 4 3 2 1

To Our Readers: We have done our best to make sure all Internet Addresses in this book were active and appropriate when we went to press. However, the author and the publisher have no control over and assume no liability for the material available on those Internet sites or on other Web sites they may link to. Any comments or suggestions can be sent by e-mail to comments@enslow.com or to the address on the back cover.

Photo Credits: AnimAlu Productions, Jet Propulsion Laboratory, NASA, p. 11; Background painting, "A cocoon nebula, perhaps the primordial solar nebula" by William K. Hartmann. Courtesy of UCLA, p. 40; M.R. Combi (The University of Michigan) and NASA, p. 17; Courtesy of the NEAR Project (JHU/APL), p. 37; Don Davis/NASA, p. 6; Don Dixon, p. 16; Hubble Space Telescope Comet Team, NASA, p. 36 (inset); K. Jobse, P. Jenniskens, and NASA Ames Research Center, p. 19; Library of Congress, p. 27; W. Liller (International Halley Watch), National Space Science Data Center, p. 29; Lunar and Planetary Institute, pp. 13, 14; H. Mikuz, Crni Vrh Observatory, Slovenia, p. 22; S. Molau and P. Jenniskens (NASA Ames Research Center), p. 20; NASA, Jet Propulsion Laboratory, pp. 4 (bottom), 23, 34, 38; National Oceanic and Atmospheric Administration, p. 25; National Space Science Data Center/European Space Agency, p. 31; © Rick Scott and Joe Orman, pp. 4 (top), 18; University of Maryland, Jet Propulsion Laboratory, California Institute of Technology, and Ball Aerospace and Technologies Corp., p. 42; U.S. Geological Survey, p. 8; U.S. Geological Survey, Flagstaff, p. 33; H.A. Weaver and T.E. Smith (Space Telescope Science Institute), J.T. Trauger and R.W. Evans (Jet Propulsion Laboratory), NASA, p. 36.

Cover Photo: H. Mikuz, Crni Vrh Observatory, Slovenia (comet); NASA/JPL (asteroid); Raghvendra Sahai and John Trauger (JPL), the WFPC2 science team, NASA, and AURA/STScI (background).

Cover photo: Comet Hyakutake and asteroid 951 Gaspra.

CONTENTS

1 The Threat from Space 5

2 Space Debris 9

3 History and Discovery of Comets 24

4 Exploring Comets and Asteroids 30

5 Comets and Asteroids: Up Close and Too Close 39

Chapter Notes 44

Glossary 46

Further Reading 47

Index 48

The comet Hyakutake and the asteroid Ida

1

The Threat from Space

On January 7, 2002, an asteroid wider than three football fields nearly hit Earth.

It came within 520,000 miles (837,000 kilometers), more than twice the distance from the Moon to Earth. Yet scientists considered this miss of more than half a million miles to be a close call. If the thousand-foot-wide asteroid had struck Earth, its impact would have been devastating.

"It would be well beyond all the atomic weapons in the U.S. and Russian arsenals," said scientist Don Yeomans of NASA's Near-Earth Object Program. "If it hit in the ocean near a continent, all the cities on that coast would be destroyed."[1]

Like other asteroids, the asteroid called 2001 YB5 is a

Comets and Asteroids—Ice and Rocks in Space

giant rock in space. If it had collided with Earth, it would have struck our planet at a speed of about 68,000 miles per hour (110,000 kilometers per hour).

"It's a fairly substantial rock," said Jay Tate of the Spaceguard Centre in Wales, Great Britain. "If it had hit us at that sort of speed, you would be taking out a medium-size country." Tate expected an area the size of Texas would be damaged in the shockwave and fires resulting from such an impact.[2]

Scientists are almost certain that asteroids like 2001 YB5, and much larger ones, have hit Earth before. Along Mexico's Yucatán Peninsula, scientists have gathered evidence of a giant asteroid impact that occurred 65 million years ago. The destruction caused by this impact may have led to the extinction of many forms of life on Earth at that time, including the dinosaurs.

This artist depiction shows an asteroid impacting Earth. Scientists believe that a large asteroid hit Earth 65 million years ago. The effects of this impact may have caused the extinction of many animals.

The Threat from Space

Earth has also likely been struck many times by comets. While asteroids are like giant mountains of rock in space, comets are like enormous dirty snowballs. Because they contain great amounts of ice, comets may have helped make Earth what it is today. Some scientists believe that comets, early in Earth's history, may have delivered the water that makes up our oceans. Although such impacts may have helped make Earth the life-filled planet that it is today, a comet impact now would not be welcome.

The impact of a comet or asteroid between half and one mile wide would cause problems around the globe. The dust thrown into the atmosphere would change the climate enough to cause crop failures. This lack of food would starve some populations and disrupt the world's economy. Larger impacts could dramatically alter Earth's climate for years. An impact could throw so much dust and debris into the atmosphere that little sunlight could get through. Without sunlight, much of the life on our planet would die.

Tracking down the comets and asteroids that might hit Earth is not the only reason scientists study them. Scientists can learn about the history of our solar system by studying comets and asteroids.

Astronomer Carolyn Shoemaker has discovered many comets. But discovering the comets is just the beginning. What is important is what astronomers can learn from them.

Comets and Asteroids—Ice and Rocks in Space

Planetary astronomer Carolyn Shoemaker is the most successful comet hunter. She has found more than eight hundred asteroids and thirty-two comets.

"Comets are the wild card when we consider the potential for impact on Earth," Shoemaker said. "It is necessary to learn more if we wish to defend our planet. We need to know a great deal more about their structure—are they solid bodies emitting gas and dust or are they . . . flying snowballs easily broken apart, or are they all shades in between? Are some asteroids really extinct comets in which the action has been shut off? Could comets provide a source of water for space travelers? Did comets bring life to Earth or the nutrients for life?"[3]

It has taken centuries of observation and study to learn the nature of comets and asteroids. With their sudden appearance and long tail streaming across the sky, comets have always grabbed people's attention. On the other hand, asteroids took longer for astronomers to discover because they are not bright or noticeable in the sky without a telescope.

2

Space Debris

Unlike comets that had been observed by humans for thousands of years, asteroids were not discovered until 1801. They had avoided detection by astronomers until then because they are small and dim. Galileo had invented the astronomical telescope in 1609, and people have been making stronger and stronger ones since then. By 1801, telescopes were just strong enough to make asteroids visible. But there is one important quality that comets and asteroids have in common. They are both bits of space debris.

Cosmic Leftovers

If we looked at all the things that exist in our solar system, such as the Sun, the planets, and moons, we

Comets and Asteroids—Ice and Rocks in Space

might consider both comets and asteroids as nothing more than space debris. In a way, that is what they really are. Comets and asteroids are the leftover bits of material that did not become part of the Sun, the planets, or the moons.

Scientists believe that more than 4 billion years ago, a huge cloud of gas and dust formed in space. Gravity and other forces caused the gas and dust to collect together at the center of the cloud. The pressure and heat within this mass of material eventually became so great that nuclear reactions took place. The Sun was born.

The Sun, our solar system's star, was at the center of the cloud. The gas and dust still swirling around the Sun was coming together in clumps at different distances from it. Slowly these clumps gathered up more material. The ones that formed closer to the heat of the Sun became planets made of rock. Today we know these planets as Mercury, Venus, Earth, and Mars. Those farther from the Sun's heat, where frozen gases could still exist, became giant planets made of gases. These planets are Jupiter, Saturn, Uranus, and Neptune.

Far beyond the region of the gas planets is Pluto, which seems to be made of rock and ice. It is very different from the other planets. In fact, some scientists think it should be classified as an asteroid or a dormant comet. So far, it is still classified as a planet.[1]

Some of the rocky material left over from the solar

Space Debris

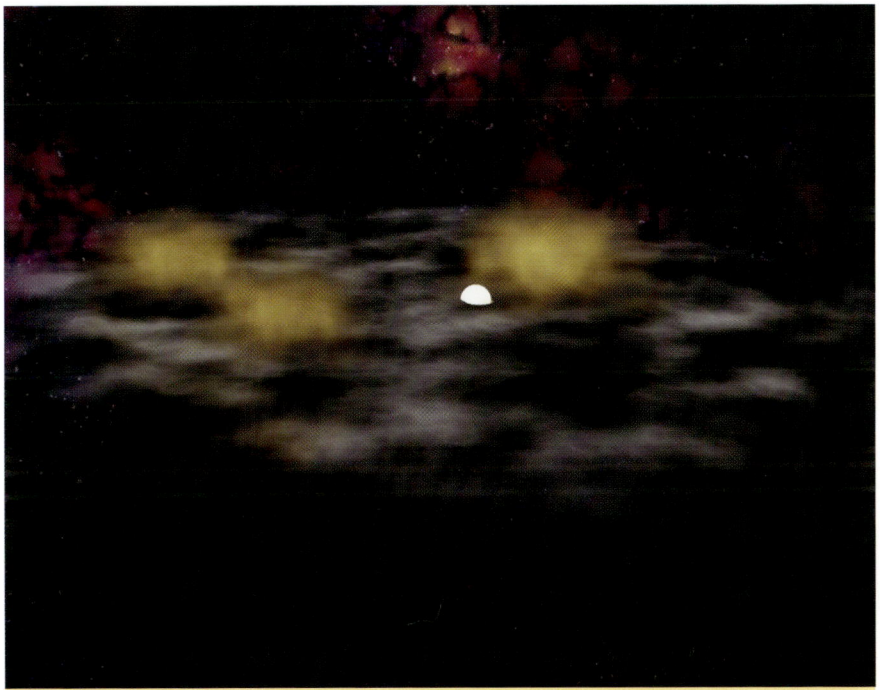

During the formation of our solar system, gases and dust formed in space. Comets and asteroids are the bits of matter that did not become the Sun or the planets.

system's formation was caught in the orbit of planets. These became some of the moons of those planets.

Dust, rock, and ice that had not become part of any of the planets or their moons still remained. These bits of leftover debris are the asteroids and comets.

What Is an Asteroid?

Asteroids are sometimes called "minor planets" because they orbit the Sun but are not large enough to be considered a planet. They are giant rocks in space, from

Comets and Asteroids—Ice and Rocks in Space

as small as a boulder to as big as hundreds of miles wide. On January 1, 1801, Sicilian astronomer Guiseppe Piazzi discovered the minor planet Ceres. At 579 miles (931 kilometers) wide, it was much smaller than the smallest known planet at that time, Mercury. Within a few years, other astronomers discovered similar objects orbiting in the area between Mars and Jupiter. These were the asteroids Pallas, Juno, and Vesta. Astronomers had discovered the asteroid belt.[2]

Asteroids orbit the Sun all over our solar system, but many thousands of asteroids are concentrated in the asteroid belt. But why is there an asteroid belt? And why does it exist between Mars and Jupiter?

During the solar system's formation, it is believed, chunks of rock came together elsewhere in the solar system and formed planets. But in the region known as the asteroid belt, the gravitational effects of the giant planet Jupiter acted on the orbits of these asteroids. These gravitational effects pushed them into tilted, more oval orbits, which caused the asteroids to collide with each other instead of forming into a giant clump that might have become a planet.[3]

The same gravitational effects exist today. The gravity of Jupiter keeps these asteroids from clumping together into larger bodies. This gravity has forced some asteroids out of their orbits in the asteroid belt and sent them inward toward the inner solar system. These are the asteroids that scientists are interested in tracking, in

Space Debris

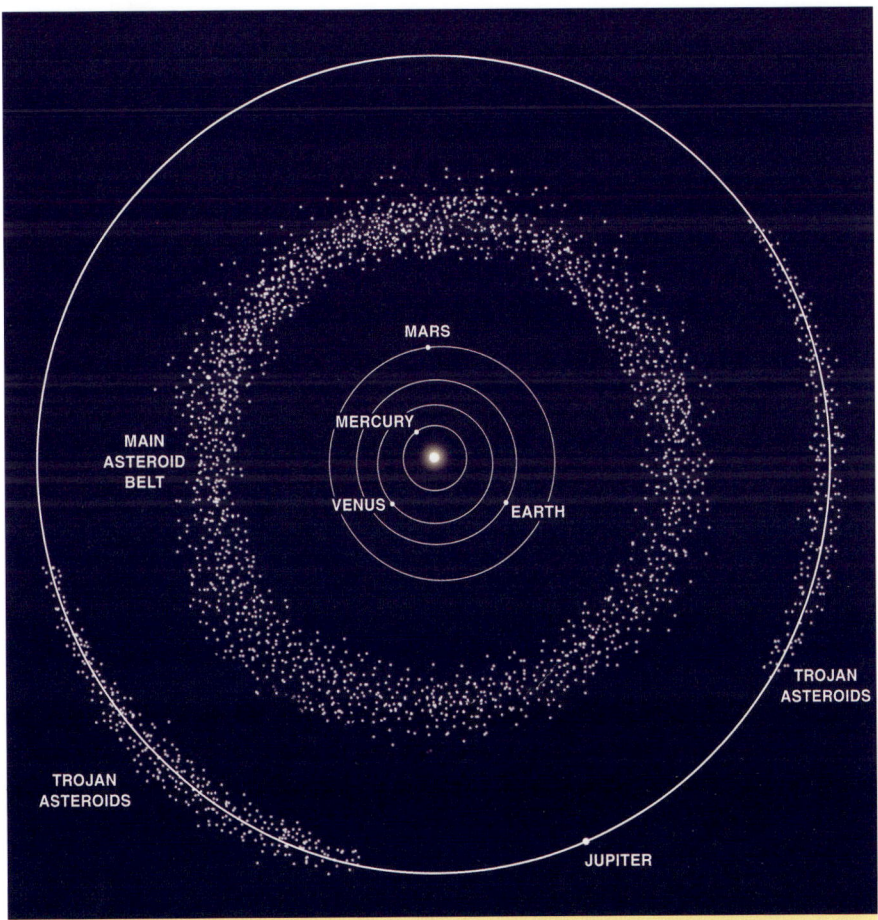

The asteroid belt lies in the region between Mars and Jupiter.

case their orbit someday carries them into a collision with Earth.

A Comet Is Born

Although humans were aware of the existence of comets long before the discovery of asteroids, there is much that scientists do not yet know about the origin of comets.

Comets and Asteroids—Ice and Rocks in Space

Currently, scientists believe that comets formed in the region of the early solar system with the giant gas planets. At this great distance from the Sun, the dust grains of the early solar system were coated with ice, which was never melted away by solar energy. The ice-coated grains collected together over millions of years into the building blocks of countless comets.

Scientists think that the gravitational effects of Uranus and Neptune gradually sent these objects farther out into the solar system, far beyond the orbit of Pluto. These objects are believed to exist now in what is like a round shell around our solar system that scientists call the Oort cloud. It is named after Dutch astronomer Jan Oort. He was the first person to think that a cloud of comets existed around our solar system, about 50,000 to 150,000 times farther from the Sun than Earth.[4]

Oort based his theory on calculations of numerous comet orbits. The orbits showed that these comets, at their most distant point, must originate from an area about

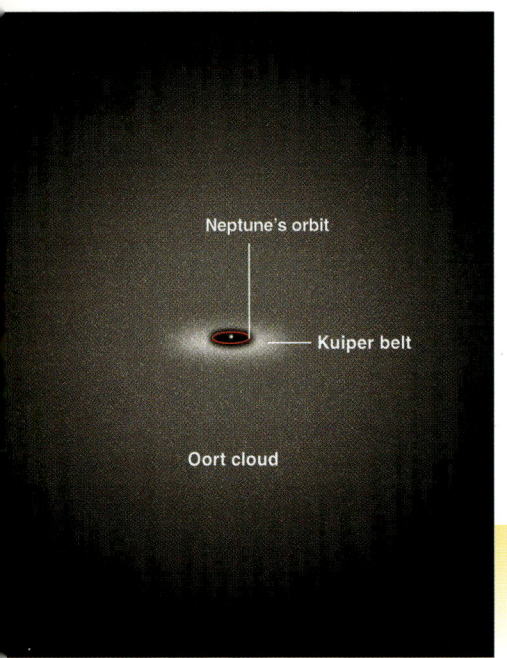

The Oort cloud is an area that circles our solar system. It contains many comets.

Space Debris

50,000 times Earth's distance from the Sun, but not farther. In other words, the comets all came from a distant area around our solar system, not from anywhere outside it, such as from another star. Today's scientists have reconsidered Oort's calculations and now believe the cloud might be as far as three times that distance.

Over millions of years, the gravitational effects of passing stars have sent many of these Oort cloud comets inward toward the Sun. Astronomers have observed many of these comets and calculated their orbits of the Sun to be many thousands of years. Oort estimated that there might be 190 billion of these comets.[5]

Another home of comets is the Kuiper Belt. It is named after American astronomer Gerard Kuiper, who in 1951 believed that a disk-shaped region of icy objects and comets existed just outside the orbit of Neptune. The Kuiper Belt is the source of all short-period comets—comets that make a complete orbit of the Sun in less than two hundred years.

What Makes Up a Comet?

Comets are made of ice and dust. The Sun's energy greatly affects them. During the time that the comet is orbiting far from the Sun, the comet is incredibly dim or invisible in our telescopes. Comets give off no visible light of their own. They can only be seen when their orbit carries them closer to the Sun. The Sun's energy

Comets and Asteroids—Ice and Rocks in Space

The Kuiper Belt, pictured here in blue, is home to comets that orbit the Sun in less than two hundred years. The Kuiper Belt lies just outside the orbit of Neptune. As shown, the orbit of Pluto extends into this Kuiper Belt.

causes bits of dust and vapor to rise from the comet, forming a visible tail of debris.

The three main parts of comets are the nucleus, the coma, and the tail. The nucleus is the solid part of the comet. It is mainly a mixture of ice, rocks, and dust. The coma—which means "hair"—is the cloud of gas and dust that rises from the nucleus and surrounds it as it nears the Sun. This happens when the comet gets within about 280 million miles (450 million kilometers) from the Sun—about three times Earth's distance from the Sun. At that distance, heat from the Sun warms the comet's water ice and causes it to sublimate, or turn directly from a solid to a gas without passing through the liquid state. Other gases and dust particles begin to rise as well. These

Space Debris

particles flow away from the nucleus at speeds averaging about 1,000 miles per hour (1,600 kilometers per hour).[6]

Observations by the Hubble Space Telescope of comet Hale-Bopp in 1998 showed that the nucleus of Hale-Bopp may have been 16 to 26 miles (26 to 42 kilometers) wide. Hale-Bopp is a very large comet.

As the comet moves nearer the Sun, the coma of gases and dust surrounding the nucleus becomes enormous. Most have been observed ranging from about 60,000 to 600,000 miles (100,000 to 1 million kilometers) in diameter. In comparison, Earth is only about 8,000 miles (13,000 kilometers) in diameter.

Despite the size of a comet's coma, three spacecraft were able to penetrate the coma of Halley's comet during

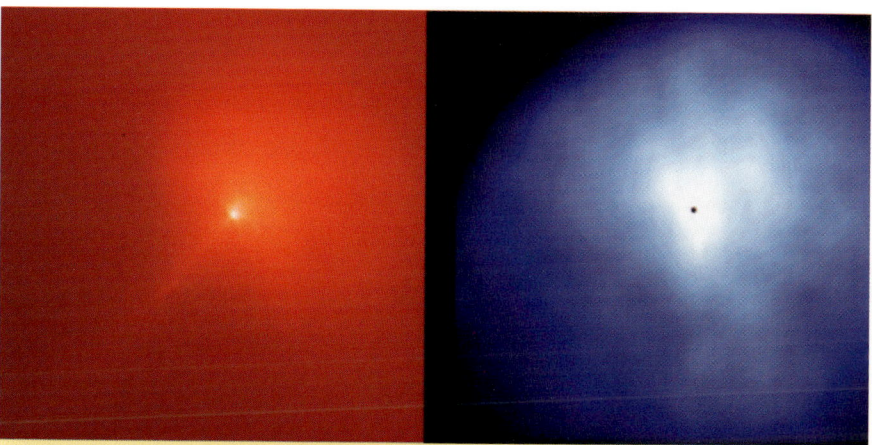

A coma of dust and gases surrounds the nucleus of a comet. This comet, Comet Hyakutake, was photographed by the Hubble Space Telescope in 1996. The red image shows dust particles in the inner coma. The blue image shows radiation from hydrogen atoms in the inner coma.

Comets and Asteroids—Ice and Rocks in Space

its passage in 1986. The European Space Agency's unmanned *Giotto* spacecraft and the Soviet Union's *Vega 1* and *Vega 2* spacecraft flew inside the coma to record images of the nucleus. Instruments aboard the three spacecraft finally confirmed the theory that comets were like dirty snowballs.

"They are the kind of snowball that . . . is . . . full of sand and rocks and other stuff," said NASA's Tom Duxbury, an investigator on the *Stardust* spacecraft, which will visit comet Wild 2 in 2004. The three spacecraft at Halley's comet detected an abundance of water, gases, and dust grains rich in the elements that scientists believe would have been present at the formation of the solar system.[7]

A Comet's Tail

A comet's tail is its most spectacular feature. These tails of gas and dust can typically stretch across space for tens of million of miles. Scientists have observed a few that have trailed from their nucleus for as far as 90 million miles (144 million kilometers), roughly equal to the distance between Earth and the Sun. In 1996, comet Hyakutake's tail was

This is comet Hyakutake in 1996. Its tail was about 350 million miles (563 million kilometers) long.

Space Debris

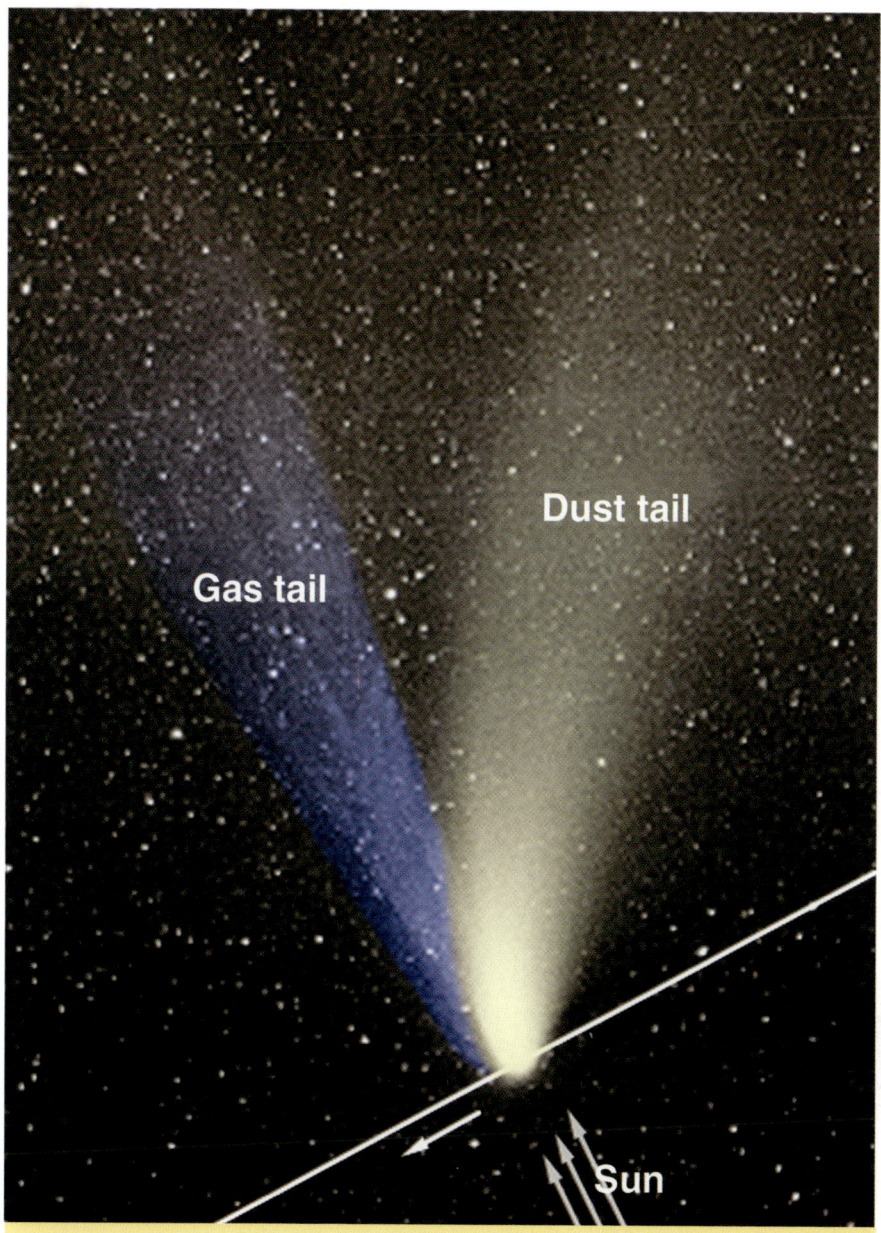

Comets have two tails. The gas tail appears blue; the dust tail appears a dim yellow.

Comets and Asteroids—Ice and Rocks in Space

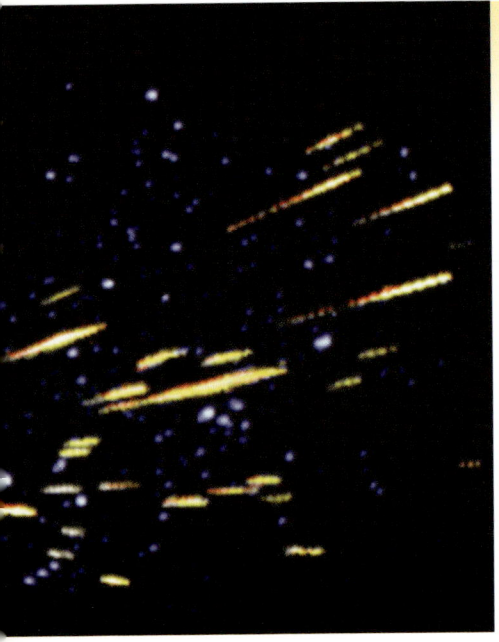

A meteor shower, like the Perseid meteor shower shown here, happens when Earth passes through comet debris.

measured to be an incredible 350 million miles (563 million kilometers) long, more than twice the distance from the Sun to the planet Mars.

A comet actually has more than one tail. One tail is the dust tail, with bits of dust so small they are like particles of smoke. These particles flow from the nucleus in a direction away from the Sun. Solar radiation causes the comet's ices and dust to spew out of cracks in the nucleus. A constant outward flow of the Sun's particles, called the solar wind, pushes the comet's dust particles away from the comet's nucleus, causing it to form this dusty trailing tail. The dust tail appears a dim yellow to our eyes, because of the way it reflects sunlight.

The trails of dust left in space by comets are what cause the meteor showers or "falling stars" we see flashing across the night sky. Most meteoroids are no larger than a grain of sand. As the particle travels into Earth's atmosphere, friction with air molecules heats the particle and vaporizes it, causing a flash of light. When

Space Debris

Earth passes directly through the dusty trail left behind by a comet in space, we experience a meteor shower.

The second tail trailing from a comet is the gas tail. It is a tail of ionized gases, or plasma, carried away from the nucleus in rays or streamers by solar magnetic fields. Gas tails appear blue because ions and carbon monoxide within the tail glow blue in sunlight.[8]

These features of a comet are the most eye-catching, but they are not the largest feature. By far the largest feature of a comet is the hydrogen cloud that surrounds it. This cloud is invisible to human eyes. In 1970, scientists studied comets Tago-Sato-Kosaka and Bennett at special wavelengths of light. They discovered huge clouds of hydrogen atoms around the comets. These clouds are so huge, they cover an area about the size of the Sun, more than 800,000 miles (1.3 million kilometers) across. Scientists believe the clouds are created by the Sun's interaction with water molecules in the comets' comas.

Telescopes on Earth have allowed astronomers to learn much about comets and asteroids. Asteroids are difficult to study by telescope because they are so small and so distant. Comets passing closer to Earth are easier to observe by telescope, but then they pass out of sight again.

Comets

Estimated number of comets in our solar system
About 190 billion

Composition of comets
Ice, rock, and dust

Where comets come from
The Kuiper Belt, beyond the orbit of Neptune;
The Oort cloud, far beyond the orbit of Pluto

Most famous comet
Halley's comet, which passes close to Earth every 76 years

Largest observed comet
Comet Hale-Bopp, with a nucleus 16 to 26 miles (26 to 42 kilometers) wide

Longest observed comet tail
Comet Hyakutake at 350 million miles (570 million kilometers)

Only comet hit ever witnessed
Comet Shoemaker-Levy 9's impact into Jupiter in 1994

Comets visited by spacecraft
Halley's comet, comet Borrelly

Asteroids

Estimated number of asteroids in our solar system
About 1 million

First asteroid discovered
Ceres, on January 1, 1801

Largest known asteroid
Asteroid 2001 KX76, at about 750 miles (1,200 kilometers) wide

Location of asteroids
All over the solar system, but hundreds of thousands are concentrated in the asteroid belt between the orbits of Mars and Jupiter

Brightest asteroid
Vesta: under excellent viewing conditions, it is the only asteroid in the asteroid belt visible to the unaided eye

Asteroids visited by spacecraft
951 Gaspra, 433 Eros, 253 Mathilde, 243 Ida

Number of asteroids considered "potentially hazardous" for striking Earth
88

3

History and Discovery of Comets

To people living in ancient times, the night sky changed predictably. People noticed how the Moon moved slowly across the sky, gradually changing its shape over a period of weeks. They noticed that the patterns of stars slowly progressed across the sky, moving from east to west, appearing again at the same time every year. It was always the same. Nothing new ever appeared in the sky.

One night, something new *did* appear. It was white and had a dusty-looking tail streaming behind it. Seeing this new object in the heavens was unsettling. People believed that such an event had to be the work of the gods. Perhaps it was a sign from the gods that something terrible was going to happen. The comet's sudden appearance created fear.

History and Discovery of Comets

The Roman writer Pliny, during a time of civil war in Rome, wrote, "We have in the war between Caesar and Pompey an example of the terrible effects which follow the apparition of a comet . . . that fearful star which overthrows the powers of Earth."[1]

People in the ancient world reacted this way many times to the appearance of comets. Science and mathematics gradually helped astronomers understand the nature of these visitors in the sky. The many appearances of one very famous comet will help to tell this story.

Historically, people viewed comets as a sign of death and destruction, as shown in this antique drawing.

Comets and Asteroids—Ice and Rocks in Space

The Visits of Halley's Comet

There have been many comets with orbits that carried them through our part of the solar system. One comet—Halley's comet—has passed our way many times.

The earliest recorded observations of Halley's comet go back to the Chinese records of 240 B.C. Over the centuries, its most stunning visits were recorded in great works of art. Its appearance in 1066 occurred at the same time that the Norman army invaded England. English King Harold II saw the comet as an evil omen. He died during the invasion that same year.

The comet made other memorable appearances in 1301, 1531, 1607, and 1682. No one knew what these and other comets were. How they moved and what they were made of was a mystery. They were not like meteors that shot across the sky and disappeared. Comets appeared, grew gradually brighter over several weeks or months, then slowly dimmed and went away.

They also moved across the sky, but slowly. Each night the comet appeared in a different position against the background of stars, creeping across the sky night after night.

English astronomer Edmond Halley was very interested in their movements. In the mid-1700s, he began studying the observations that previous astronomers had made of comets' appearances, movements, and positions. To these records, Halley applied Sir Isaac Newton's new theories of gravitation

History and Discovery of Comets

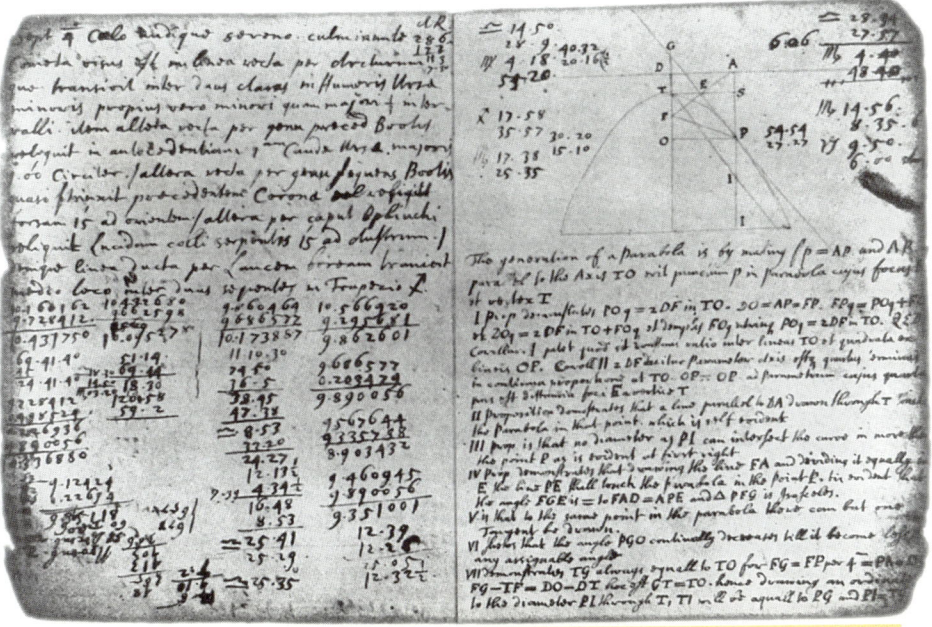

These are notebook pages from Halley's original observations of the comet in 1682. With careful calculations, he figured out the orbit of the comet that was eventually named after him.

and planetary motion. Using these theories, Halley calculated orbits for many of the best and most accurately observed comets in the historical records.

Halley discovered that the orbits of the comets of 1531, 1607, and 1682 were almost exactly the same. This discovery led him to believe that this was the same comet returning along its orbital path about every seventy-six years. He predicted that the comet would return in 1758. Halley did not live to see if his prediction was right. He died in 1742.[2]

Comets and Asteroids—Ice and Rocks in Space

On Christmas Day 1758, a German amateur astronomer named Johann Palitzsch spotted the returning comet in his telescope. Through the last week of December 1758 and the early part of 1759, the comet displayed itself across Earth's night sky, just as Halley had predicted. The comet was named Halley's comet in his honor. Astronomers officially designated the comet as 1P, meaning that it was the first known periodic comet.[3]

What is a periodic comet? Are there other kinds of comets?

Halley helped answer those questions as well.

The Orbits of Comets

Halley was able to predict the comet's 1758 appearance by calculating the comet's orbit. He discovered the comet's orbit was very elliptical, or oval. The orbits of all the planets around the Sun are also elliptical, especially the orbit of distant Pluto. But the orbits of the planets are only slightly elliptical compared to the orbits of Halley's and other comets.

Halley discovered that the comet's orbit around the Sun carried it as close to the Sun as 50 million miles (81 million kilometers) and as far away as 3.3 billion miles (5.3 billion kilometers). This meant that Halley's orbital path carried it inside the orbit of Venus at its closest, and out beyond the orbit of Neptune at its farthest. Because its seventy-six-year orbit is known, it is called a periodic

History and Discovery of Comets

Halley's comet orbits the Sun about every seventy-six years.

comet. There are many other periodic comets observed and recorded by astronomers, but they are not as bright and spectacular as Halley's comet.[4]

Halley's comet is the only famous comet that has an orbital period of less than five hundred years. The rest of the great comets that have captured the world's attention have much longer orbital periods. This means their orbits are much more elliptical than Halley's comet, and that the most distant point of their orbits carries them far out beyond the orbit of Pluto. Based on observations of these comets, astronomers have determined that the orbital period of some of these comets is measured in thousands of years.

To learn more about comets and asteroids, scientists have sent spacecraft to visit them.

4

Exploring Comets and Asteroids

Scientists were ready when Halley's comet came through Earth's part of the solar system in 1986. The European Space Agency launched the *Giotto* spacecraft, the Soviet Union launched the *Vega 1* and *Vega 2* probes, and Japan sent the *Sakigake* and *Suisei* spacecraft. All of them were sent to meet the comet in March 1986.

The *Vega 1* probe (on March 6) and the *Vega 2* probe (on March 9) photographed the nucleus of a comet for the first time. *Vega 1* came within 5,600 miles (9,000 kilometers) of the nucleus, and *Vega 2* within 5,100 miles (8,200 kilometers). It was difficult to distinguish the solid surface of the nucleus from the cloud of gas and dust around it. The images were analyzed and coded by

Exploring Comets and Asteroids

the brightness of light so that the approximate size of the nucleus could be found.[1]

Better images of Halley's nucleus were taken by the *Giotto* spacecraft on March 14, 1986. *Giotto* came within 1,692 miles (2,730 kilometers) of Halley's nucleus, recording images with its multicolor camera. The images showed the nucleus to be about 9 miles (15 kilometers) wide and potato-shaped, not round or spherical as previously thought.

Halley's nucleus is extremely dark. Analysis of the images from *Giotto* showed that the dusty crust of the nucleus is actually darker than coal.

The images also showed that the nucleus was rotating, and that dust and gas were not flowing uniformly from all surfaces of the nucleus. Some parts of the surface were releasing material while others were not. Scientists believe particles were leaving the nucleus where the dark crust of the comet was absent. The absence of this crust exposed the interior ices

The Giotto *spacecraft photographed the nucleus of Halley's comet on March 14, 1986. The image allowed scientists to see that the nucleus was potato-shaped, not round.*

Comets and Asteroids—Ice and Rocks in Space

to sunlight, which caused the ices to rise into space as gas. Only about 10 percent of the nucleus surface area was active when *Giotto* and the two Vega spacecraft flew by.[2] This rotation of the nucleus, and the nonuniform rising of gas and dust from its surface, explained why comets sometimes appear to have irregular streamers of material flowing away from them.

Galileo Peeks at an Asteroid

The *Galileo* spacecraft was launched in October 1989. Its main mission was to conduct a four-year study of Jupiter and its moons. But on its journey to the giant planet, *Galileo* made the first flyby of an asteroid and observed the most spectacular comet event ever witnessed.

On October 29, 1991, as *Galileo* passed through the asteroid belt, the spacecraft recorded the first close-up image of an asteroid. From about 10,000 miles (16,000 kilometers), *Galileo* revealed the cratered surface of asteroid 951 Gaspra. It was about 12 miles (20 kilometers) wide and shaped like a potato. Its shape suggested that it was a fragment of a larger asteroid that was broken apart by collisions. Perhaps it was still spinning from one of those collisions. *Galileo* observed that this asteroid rotated once every seven hours.

In 1993, the spacecraft passed close to another asteroid called 243 Ida. The asteroid is 36 miles (58

Exploring Comets and Asteroids

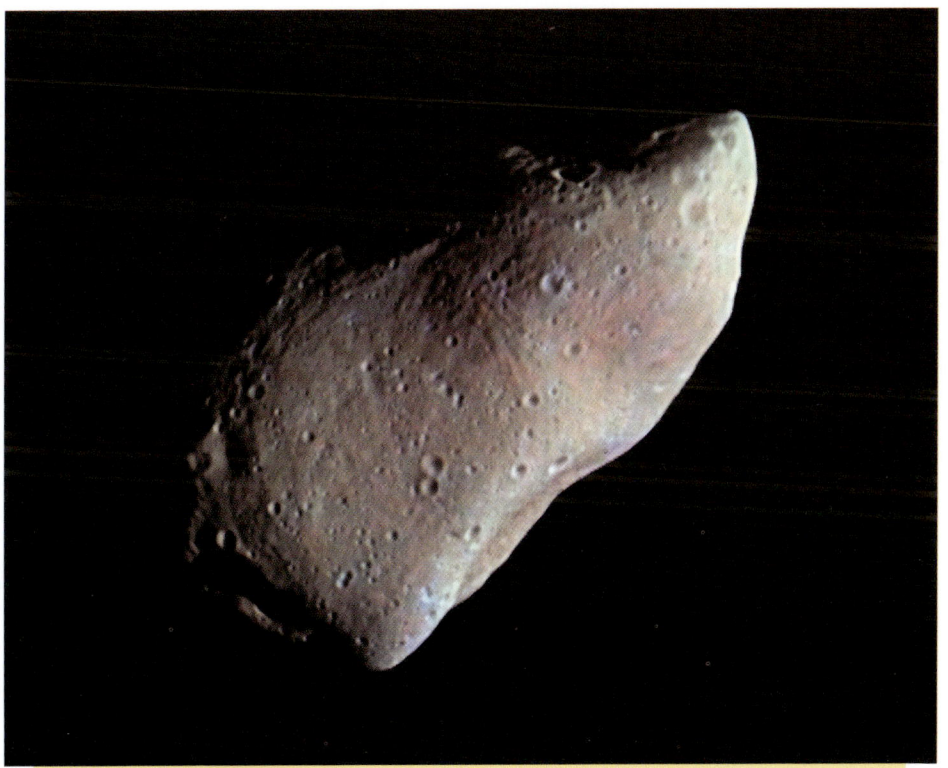

While traveling through the asteroid belt, the Galileo spacecraft photographed the asteroid 951 Gaspra. This picture is a combination of two images and is the first close-up image of an asteroid. The Sun is shining from the right.

kilometers) wide and shaped similar to Gaspra. But *Galileo*'s great accomplishment this time was the discovery of the first moon in orbit around an asteroid. Later named Dactyl, the tiny moon was only about 1 mile (1.6 kilometers) wide, yet it was covered with craters, some wider than a football field.[3]

As *Galileo* continued toward its mission at Jupiter,

Comets and Asteroids—Ice and Rocks in Space

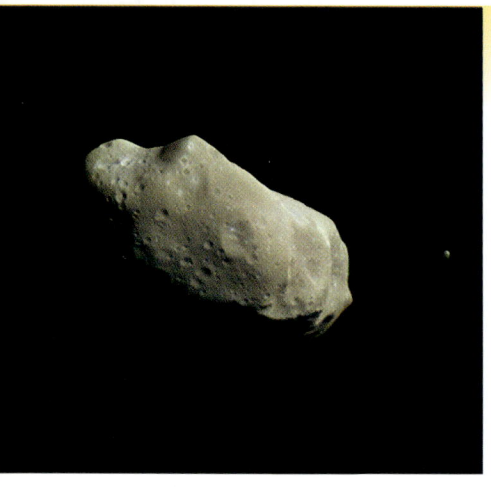

The asteroid 243 Ida has a moon that orbits it. The moon's name is Dactyl.

something amazing happened. The spacecraft was still 121 million miles (196 million kilometers) from Jupiter when fragments of comet Shoemaker-Levy 9 smashed into the planet.

Observing Shoemaker-Levy 9

Astronomers Eugene and Carolyn Shoemaker and David Levy had discovered the comet in March 1993. It had been torn apart and was in several pieces with tails streaming from them. Scientists later calculated that the comet's orbit had recently carried it within 13,000 miles (21,000 kilometers) of Jupiter. At that distance, the mighty gravity of Jupiter had torn the comet apart. To the excitement of astronomers around the world, when the comet again passed Jupiter, the gravity would drag the fragments straight into the planet. *Galileo* was on its way to Jupiter and was in a great position to observe the event.[4]

Shoemaker-Levy 9 was in twenty-one fragments when it arrived at Jupiter in July 1994. *Galileo* recorded some amazing images of the huge disturbances the comet impacts made in the planet's cloud bands. The

Exploring Comets and Asteroids

spacecraft also recorded the impacts with a spectrometer, an instrument for analyzing light. Studying the light emitted during the collisions helped tell scientists the specific elements that made up the comet.

The Hubble Space Telescope, which orbits above Earth's atmosphere, took even more pictures of the comet's impact with Jupiter. Some of the more spectacular images showed plumes rising above the planet following a comet fragment's impact. One of the plumes rose more than 2,000 miles (3,200 kilometers). The dark cloud the impact created in the planet's atmosphere was the size of Earth. Despite the strong winds in Jupiter's atmosphere, the dark spots created by the four largest impacts lasted almost a year.[5]

Comet Shoemaker-Levy 9's impact with Jupiter was the most violent event in our solar system's recorded history. There would be human survivors if a comet similar to it ever struck Earth, but the impacts, and the fires and the dust-clouded atmosphere that would follow for months after, would destroy our civilization. We are fortunate it struck Jupiter instead of Earth.

Other Asteroid and Comet Encounters

The *Near Earth Asteroid Rendezvous (NEAR) Shoemaker* spacecraft, named in honor of Eugene Shoemaker, who died in 1997, encountered asteroid 253 Mathilde in June 1997. Images from this passage showed that 253 Mathilde was likely a collection of rocky fragments

Comets and Asteroids—Ice and Rocks in Space

On May 17, 1994, the icy fragments of the comet Shoemaker-Levy 9 headed toward Jupiter. On July 21, 1994, the Hubble Space Telescope photographed the impact regions on Jupiter (inset). The fragments are labeled with letters. This image was taken only hours after fragment R hit Jupiter's atmosphere.

Exploring Comets and Asteroids

Before landing on asteroid 433 Eros, the Near Earth Asteroid Rendezvous (NEAR) Shoemaker spacecraft took these images of the asteroid. The spacecraft found out that the asteroid was made of solid rock.

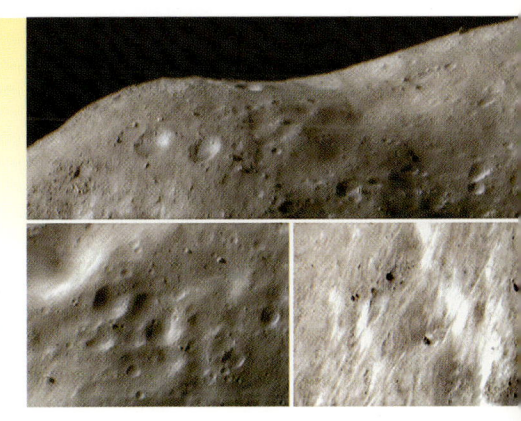

rather than a single giant piece of rock.

The *NEAR* spacecraft then passed asteroid 433 Eros in December 1998. The spacecraft later returned to asteroid 433 Eros and went into orbit around it in February 2000. *NEAR* completely mapped the asteroid and found its composition to be quite different from 253 Mathilde's. The spacecraft's instruments indeed found that 433 Eros was a giant solid rock, unlike the rocky fragments of 253 Mathilde. The differences between these two asteroids illustrated the varied composition of asteroids and the role that collisions play in their makeup.[6]

But *NEAR*'s mission at 433 Eros was not over. On February 12, 2001, the *NEAR* spacecraft touched down on the asteroid, taking sixty-nine close-up pictures as it descended. On the asteroid's surface, the spacecraft's other instruments will help scientists determine the major elements that make up the asteroid. Determining these elements and how much of them exist on 433 Eros will help answer scientists' questions about how old the

Comets and Asteroids—Ice and Rocks in Space

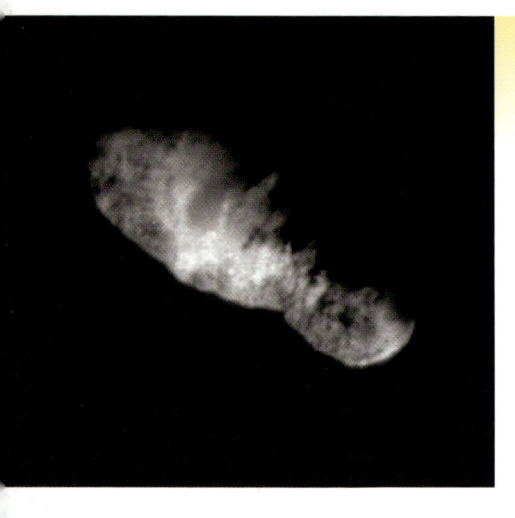

The Deep Space 1 spacecraft visited comet Borrelly and took the clearest picture ever of the nucleus of a comet.

asteroid is, and how and where in our solar system it was formed.[7]

The *Deep Space 1 (DS1)* spacecraft approached comet Borrelly in September 2001. From 1,300 miles (2,100 kilometers) away, *DS1* took the best pictures ever of a comet nucleus.

"These pictures have told us that comet nuclei are far more complex than we ever imagined," said Larry Soderblom of the *DS1* imaging team. "They have rugged terrain, smooth rolling plains, deep fractures, and very, very dark material."[8]

"The lessons learned from this cometary encounter with Borrelly will be put to good use for the coming five encounters with comets in the next decade," said Don Yeomans, manager of NASA's Near-Earth Object Program.[9]

With telescopes and spacecraft, our knowledge of comets and asteroids increases. But the search for more answers continues.

5

Comets and Asteroids: Up Close and Too Close

Scientists' efforts to learn more about comets and asteroids are just beginning. Several spacecraft will be visiting comets and asteroids in the coming years.

MUSES-C is designed to be the first spacecraft to collect asteroid surface samples and return them to Earth. It is scheduled to land on asteroid 1998 SF36 in June 2005. After making a five-month study of the asteroid's surface, *MUSES-C* will lift off from the asteroid with an asteroid surface sample and return to Earth in June 2007.

The *Dawn* mission is even more ambitious. *Dawn* is scheduled to visit two of the largest known asteroids, Ceres and Vesta. If its mission to these two residents of the asteroid belt is successful, *Dawn* will become the first

Comets and Asteroids—Ice and Rocks in Space

spacecraft to orbit two planetary bodies on a single voyage. The spacecraft is set to begin studying Vesta in July 2010 and Ceres in August 2014.

The two asteroids were chosen for the spacecraft's visits because they are different kinds of asteroids with what are believed to be vastly different histories. Scientists think Ceres was formed at a distance from the Sun where water remained as one of its components. Frost or vapor may still exist on its rocky surface, or within its minerals. Vesta, however, may have developed under hot and violent conditions. Scientists believe its

The Dawn *spacecraft will visit the asteroids Vesta and Ceres.*

Comets and Asteroids: Up Close and Too Close

interior is molten. Studying these different asteroids closely will provide more clues about the formation of our solar system.[1]

One of the most interesting comet missions is the European Space Agency's *Rosetta* mission. After encounters with asteroid 4979 Otawara in 2006 and asteroid 140 Siwa in 2008, *Rosetta* will reach comet Wirtanen in 2011.

The *Deep Impact* spacecraft is scheduled to arrive with a bang on comet Tempel 1 in 2005. The impactor section of *Deep Impact* will separate from the flyby spacecraft and create a crater the size of a football field, seven stories deep, on the comet's surface. The impactor will have cameras to provide close-up pictures of the comet nucleus before impact. The other spacecraft will record the impact and its aftermath, collecting data on how the impact affects the comet.

Seeing a Comet or Asteroid

You cannot see an asteroid with your unaided eyes, and a comet needs to move close enough to the Sun to form a coma and tail for you to see it, even with a telescope.

An experienced amateur astronomer with a quality telescope could show you an asteroid or two, but they will likely look little different to you than a dim star. As larger comets move closer to the Sun, they can of course become easily visible to the unaided eye.

You can check astronomy magazines like *Astronomy*

Comets and Asteroids—Ice and Rocks in Space

DEEP IMPACT *First Look Inside A Comet*

The Deep Impact *spacecraft will take a look inside the comet Tempel 1.*

and *Sky & Telescope* to find out about asteroids and comets that may currently be visible in the night sky. If a spectacular comet is on the way, you will probably hear about it on the television news or in the newspaper. You can also look at Web sites on the Internet that contain information about asteroids and comets.

Will a comet or asteroid ever come *too* close? Maybe. It has happened many times before in Earth's history, and quite recently.

On June 30, 1908, a terrific explosion thundered

Comets and Asteroids: Up Close and Too Close

through the Tunguska region of Siberia in Russia. The explosion leveled trees in this undeveloped part of the world for hundreds of square miles. But there was no evidence of an earthquake or a volcano. There were no nuclear bombs in 1908.

Not until fifty years later did scientists offer an explanation for what happened in Tunguska. Further evidence collected since then supports the theory that a comet or asteroid 50 to 60 yards (46 to 55 meters) wide and weighing about 1 million tons exploded in the atmosphere above Tunguska. The force of the explosion was equal to that in about one thousand nuclear bombs.[2]

It is unlikely that a large comet or asteroid will strike Earth anytime soon. But events such as the Tunguska explosion and the Shoemaker-Levy 9 impact at Jupiter prove that such events are not just part of ancient history.

Until the impact of Shoemaker-Levy 9 at Jupiter, "no one had ever seen the impact upon a planet of a comet or an asteroid," said the comet's codiscoverer Carolyn Shoemaker. "This comet was nature's grand experiment for man to learn about the structure and makeup of comets, impact dynamics . . . [and] most importantly, to gain the awareness that, yes, objects can fall from the sky and impact planets, including ours."[3]

Comets and asteroids can be a great threat to our Earthly existence, yet they may also hold many answers to how our solar system and everything in it, including us, ever came to be.

CHAPTER NOTES

Chapter 1. The Threat from Space

1. Steven Siceloff, "Earth Escapes Close Call with Massive Asteroid," *Asteroid and Comet Impact Hazards*, January 8, 2002, <http://web99.arc.nasa.gov/impact/news_detail.cfm?ID=105> (January 21, 2002).

2. The Associated Press, "Asteroid Big Enough to Raze France Zips by Earth," *Asteroid and Comet Impact Hazards*, January 8, 2002, <http://web99.arc.nasa.gov/impact/news_detail.cfm?ID=105> (January 21, 2002).

3. Jennifer Laing, "Comet Hunter," *The Universe Today*, December 11, 2001, <http://www.universetoday.com/html/articles/2001-1211a.html> (January 21, 2002).

Chapter 2. Space Debris

1. Michael E. Bakich, *The Cambridge Planetary Handbook* (New York: Cambridge University Press, 2000), p. 302.

2. Donald K. Yeomans, *Comets* (New York: John Wiley & Sons, 1991), pp. 148–149.

3. Jean Audouze and Guy Israel, eds., *The Cambridge Atlas of Astronomy* (Cambridge, England: Cambridge University Press, 1996), p. 162.

4. David H. Levy, *Comets: Creators and Destroyers* (New York: Simon & Schuster, 1998), p. 25.

5. Yeomans, pp. 317–320.

6. J. Kelly Beatty, Carolyn Collins Petersen, and Andrew Chaikin, eds., *The New Solar System* (Cambridge, Mass.: Sky Publishing Corporation, 1999), p. 327.

7. William Speed Weed, "Chasing a Comet," *Astronomy*, September 2001, p. 34.

8. Beatty, Petersen, and Chaikin, p. 325.

Chapter Notes

Chapter 3. History and Discovery of Comets
1. Patrick Moore, *The Picture History of Astronomy* (New York: Grosset & Dunlap, 1961), p. 81.
2. Donald K. Yeomans, *Comets* (New York: John Wiley & Sons, 1991), pp. 118–123.
3. Moore, p. 82.
4. J. Kelly Beatty, Carolyn Collins Petersen, and Andrew Chaikin, eds., *The New Solar System* (Cambridge, Mass.: Sky Publishing Corporation, 1999), p. 324.

Chapter 4. Exploring Comets and Asteroids
1. Jean Audouze and Guy Israel, eds., *The Cambridge Atlas of Astronomy* (Cambridge, England: Cambridge University Press, 1996), p. 234.
2. Ibid., pp. 234–235.
3. "Missions to Asteroids: Galileo," *Solar System Exploration*, February 1, 2002, <http://sse.jpl.nasa.gov/missions/ast_missns/ast-galileo.html> (March 1, 2002).
4. David H. Levy, *Impact Jupiter* (New York: Plenum Press, 1995), pp. 1–3.
5. Ibid., pp. 220–250.
6. "Near Earth Asteroid Rendezvous (NEAR)," *Near-Earth Object Program*, February 19, 2002, <http://neo.jpl.nasa.gov/missions/near.html> (February 25, 2002).
7. "Landing on Eros," *NEAR Science Update*, February 20, 2001, <http://near.jhuapl.edu/news/sci_updates/01feb20.html> (January 21, 2002).
8. Jeff Foust, "Deep Space 1 Returns Stunning Images of Comet," *Spaceflight Now*, September 25, 2001, <http://spaceflightnow.com/news/n0109/25ds1flyby/> (September 26, 2001).
9. Ibid.

Chapter 5. Asteroids and Comets: Up Close and Too Close
1. "Dawn," *Near-Earth Object Program*, February 19, 2002, <http://neo.jpl.nasa.gov/missions/near.html> (February 25, 2002).
2. C. Chyba, P. Thomas, and K. Zahnle, "The 1908 Tunguska Explosion: Atmospheric Disruption of a Stony Asteroid." *Nature*, v. 361, pp. 40–44.
3. Jennifer Laing, "Comet Hunter," *The Universe Today*, December 11, 2001, <http://www.universetoday.com/html/articles/2001-1211a.html> (January 21, 2002).

GLOSSARY

asteroid belt—An area of our solar system between the orbits of Mars and Jupiter that contains thousands of asteroids.

asteroids—Very small planets made of rock and metal. They are like giant boulders or mountains in space.

coma—The cloudy envelope of gas and dust that surrounds a comet nucleus as its orbit carries it nearer the Sun.

comet—A frozen clump of ice and dust in space that forms long streaming tails of dust and gas when orbiting near the Sun.

dust tail—The long tail of dust particles, millions of miles long, that streams from a comet as it orbits nearer the Sun.

gas tail—The stream of particles swept away from a comet nucleus by the solar wind.

Kuiper Belt—An area of comets just beyond the orbit of Neptune.

meteor—The streak of light seen in the sky when a meteoroid, sometimes a grain of dust or sand left behind by a comet, enters a planet's atmosphere.

nucleus (of a comet)—The solid, frozen part of a comet, made of ice and dust.

Oort cloud—A vast, spherical envelope of comets that surrounds our solar system beyond the orbit of Pluto.

orbit—The curved path of a body in space around another body in space, brought about by the force of gravity.

solar wind—A constant outward flow of particles from the Sun.

FURTHER READING

Books

Bortz, Fred, and Alfred Bortz. *Collision Course: Cosmic Impacts and Life on Earth.* Brookfield, Conn.: Millbrook Press, 2001.

Man, Jon. *Comets, Meteors, and Asteroids*. New York: DK Publishing, 2001.

Vogt, Gregory. *Asteroids, Comets, and Meteors*. Chatham, N.J.: Raintree Steck-Vaughn Publishers, 2000.

Internet Addresses

Hamilton, Calvin J. "Asteroid Introduction." *Views of the Solar System.* <http://www.solarviews.com/eng/asteroid.htm>.

Morrison, David. "Asteroid and Comet Impact Hazards." *NASA Ames Research Center*. <http://impact.arc.nasa.gov/index.html>.

Solar System Exploration. "Missions to Asteroids." <http://sse.jpl.nasa.gov/missions/ast_missns/ast-muses.html>.

INDEX

A
asteroid
  belt, 12, 13, 23, 32
  composition, 7, 11
  discovery, 9, 12, 24, 25
  exploration, 32–38
  impacts, 5, 6, 43
  observing, 41, 42
  origins, 10, 11
  size, 11–12, 22
asteroids
  Ceres, 12, 23, 39, 40
  433 Eros, 23, 37
  951 Gaspra, 23, 32, 33
  243 Ida, 23, 32, 33, 34
  253 Mathilde, 23, 35, 37
  Vesta, 12, 23, 39, 40–41

C
comet
  composition, 7, 11, 15, 16–20, 22
  discovery, 26, 27
  exploration, 18–36
  impacts, 7, 43
  observing, 41, 42
  old ideas about, 24, 25
  orbits, 28, 29
  origins, 10, 11, 14, 22
comets
  Bennet, 21
  Borrelly, 22, 38
  Hale-Bopp, 17, 22
  Halley's, 17, 18, 22, 26–29, 30–31
  Hyakutake, 17, 18, 20, 22
  Shoemaker-Levy 9, 22, 34, 35, 36, 43
  Tago-Sato-Kosaka, 21, 22

D
Dawn spacecraft, 39, 40
Deep Impact spacecraft, 41, 42
Deep Space 1 spacecraft, 38

G
Galileo spacecraft, 32–34
Giotto spacecraft, 18, 30–32

H
Halley, Edmond, 26–28
Hubble Space Telescope, 17, 35, 36

K
Kuiper Belt, 15, 16, 22

M
meteor showers, 20–21
MUSES-C spacecraft, 39

N
Near-Earth Object program, 5, 38
NEAR Shoemaker spacecraft, 35, 37

O
Oort Cloud, 14, 15, 22

R
Rosetta spacecraft, 41

S
Shoemaker, Carolyn, 7, 8, 34, 43
Shoemaker, Eugene, 34, 35

V
Vega spacecraft, 18, 30, 32

```
523.6 C675c
Cole, Michael D.
Comets and asteroids : ice
and rocks in space
```
STV FEB 21 2006